·KEITH LYE·ALASTAIR CAMPBELL·

Atlas in the Round

RUNNING PRESS
PHILADELPHIA · LONDON

Contents

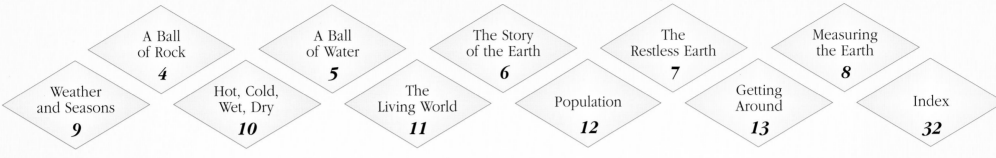
The Maps

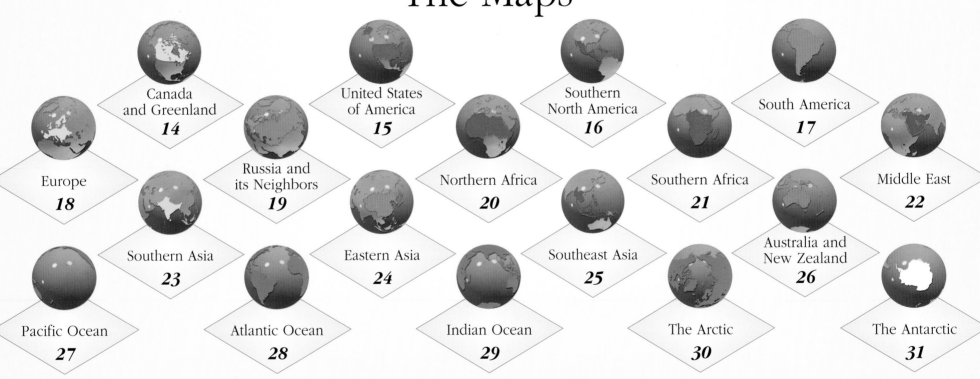
Library of Congress Cataloging-in-publication number
99-70312
ISBN 0-7624-0657-7

Art Director: Peter Bridgewater
Editorial Director: Sophie Collins
Designer: Alastair Campbell
Maps: Nicholas Rowland

This book may be ordered by mail from the publisher.
Please include $2.50 for postage and handling.
But try your bookstore first!
Running Press Book Publishers
125 South Twenty-second Street
Philadelphia, Pennsylvania 19103-4399
Visit us on the web!
www.runningpress.com

Introduction

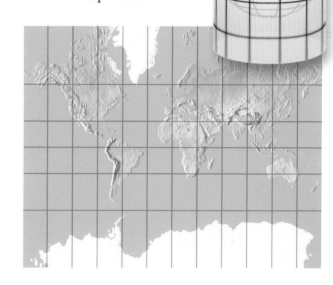

The *ATLAS IN THE ROUND* is a new and exciting book about Planet Earth. Instead of being an atlas that shows the world in flat maps, it contains views of our round world as they would appear from space. These views are like snapshots taken by astronauts from space but with all the clouds removed. The views of the world in *ATLAS IN THE ROUND* are like those you can seen on globes, which are models of our round Earth. Globes accurately show distances, areas, directions and the correct shapes of land and sea areas.

Unlike globes, flat maps of large areas can never be completely accurate. This is because it is impossible to show a round surface on a flat piece of paper. You can understand why this is so if you peel an orange. There is no way you can flatten the orange peel without breaking it up and stretching the pieces.

One way of showing the round Earth on a map is to divide the globe into segments, as shown here. But maps made in this way break up land and sea areas. This makes it hard to see the true shapes of the continents and the oceans.

Another way to make a world map is to imagine a paper cylinder wrapped around a transparent globe. If you place a light at the center of the globe, it will cast, or project, shadows of the lines on the map onto the cylinder. These shadows would form a world map. This map "projection" is accurate along the equator, where the paper touches the globe. But areas near the poles look much bigger than they really are.

This atlas provides information about Planet Earth. It also takes you on a fascinating journey around the world, showing the oceans and land as they would appear from space.

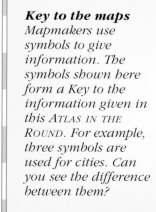

Key to the maps		
Mapmakers use symbols to give information. The symbols shown here form a Key to the information given in this ATLAS IN THE ROUND. For example, three symbols are used for cities. Can you see the difference between them?	⬤	*Capital cities*
	●	*Cities*
	●	*State/Territory capitals*
	～	*Rivers*
	～	*International borders*
	～	*State/Territory borders*
	▲	*Mountain peaks*

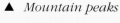

A Ball of Rock

The Earth is a sphere (ball) of rock. It is one of nine planets that travel around the Sun in the solar system. Its surface contains large land areas, called continents, and huge oceans. The oceans look smooth from space. But if we could drain off the water, as shown in the globes on this page, we would see that the ocean floor is just as uneven as the land areas, with mountains, valleys, and plains.

When the Earth first formed, around 4.6 billion years ago, hot molten (liquid) rock covered the surface. As the Earth cooled, the surface hardened to form a thin crust. Steam from volcanoes formed clouds. Rainwater from storms collected in hollows in the surface. These were the first oceans.

The Earth's hard outer layers are split, like a cracked egg, into several large blocks, called plates. The plates move around because of slow movements in the partly molten rocks below them. As a result of this, the Earth's surface is always changing.

If you cut through an apple, you will see that it has a skin that protects the fleshy interior. The apple skin is thin, like the Earth's hard crust. Just like the Earth, the apple also has a core. Between the skin and the core is the flesh of the apple. This is the equivalent of the part of the Earth that scientists call the Earth's mantle.

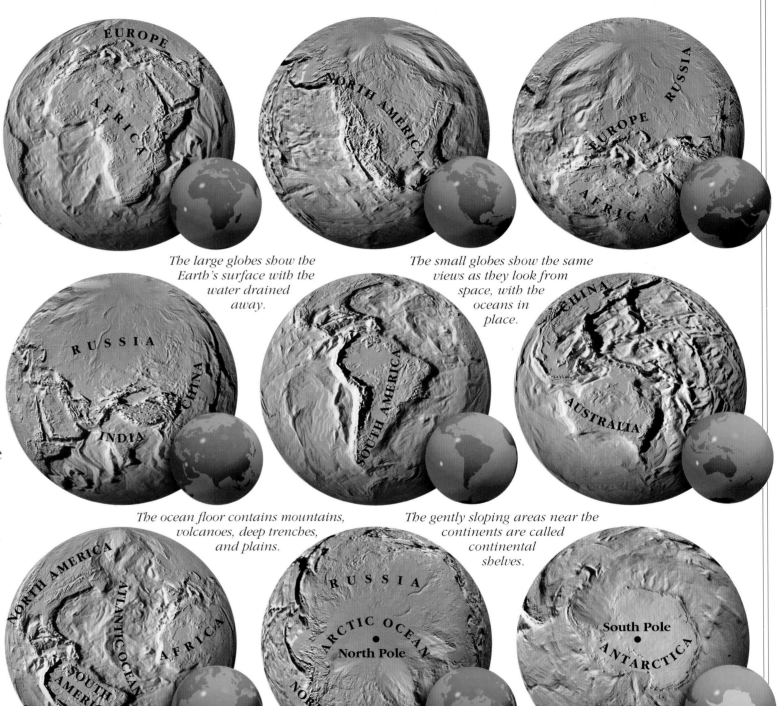

The large globes show the Earth's surface with the water drained away.

The small globes show the same views as they look from space, with the oceans in place.

The ocean floor contains mountains, volcanoes, deep trenches, and plains.

The gently sloping areas near the continents are called continental shelves.

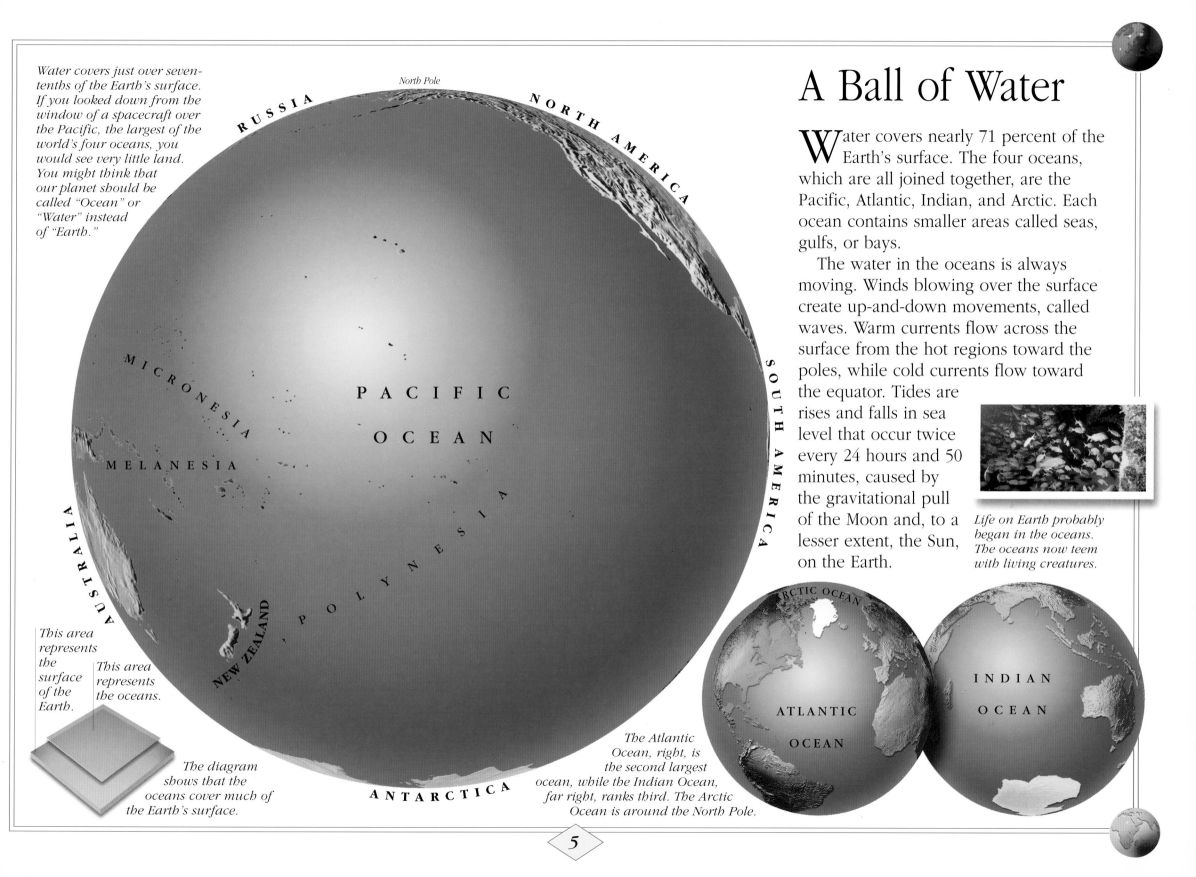

Water covers just over seven-tenths of the Earth's surface. If you looked down from the window of a spacecraft over the Pacific, the largest of the world's four oceans, you would see very little land. You might think that our planet should be called "Ocean" or "Water" instead of "Earth."

RUSSIA

NORTH AMERICA

MICRONESIA

MELANESIA

AUSTRALIA

PACIFIC

OCEAN

SOUTH AMERICA

POLYNESIA

NEW ZEALAND

ANTARCTICA

This area represents the surface of the Earth.

This area represents the oceans.

The diagram shows that the oceans cover much of the Earth's surface.

A Ball of Water

Water covers nearly 71 percent of the Earth's surface. The four oceans, which are all joined together, are the Pacific, Atlantic, Indian, and Arctic. Each ocean contains smaller areas called seas, gulfs, or bays.

The water in the oceans is always moving. Winds blowing over the surface create up-and-down movements, called waves. Warm currents flow across the surface from the hot regions toward the poles, while cold currents flow toward the equator. Tides are rises and falls in sea level that occur twice every 24 hours and 50 minutes, caused by the gravitational pull of the Moon and, to a lesser extent, the Sun, on the Earth.

Life on Earth probably began in the oceans. The oceans now teem with living creatures.

ARCTIC OCEAN

ATLANTIC

OCEAN

INDIAN

OCEAN

The Atlantic Ocean, right, is the second largest ocean, while the Indian Ocean, far right, ranks third. The Arctic Ocean is around the North Pole.

The Story of the Earth

The Earth has seen many changes in its 4.6-billion-year-long history. At first, the Earth's surface was probably ablaze with molten rocks, so no rocks formed more than about 4 billion years ago have been found. The earliest known fossils, of microscopic organisms, are around 3.5 billion years old. But amphibians, the first land animals, did not appear until between 408 and 360 million years ago. Dinosaurs lived between about 220 and 65 million years ago. Mammals became common in the last 65 million years. Modern people appeared just over 100,000 years ago.

Plate movements have changed the face of the Earth. Aliens visiting our planet around 200 million years ago would have seen only one super-continent that was called Pangaea. But in the last 180 million years, Pangaea has broken apart and the continents have moved to their present positions.

Ammonites were mollusks that once lived in the sea. They became extinct about 65 million years ago. The picture shows a fossil ammonite. It is a cast of the original animal in rock, which preserves its shape. Fossils are evidence of ancient life found in rocks. From fossil evidence, scientists have pieced together the story of how life has evolved on Earth.

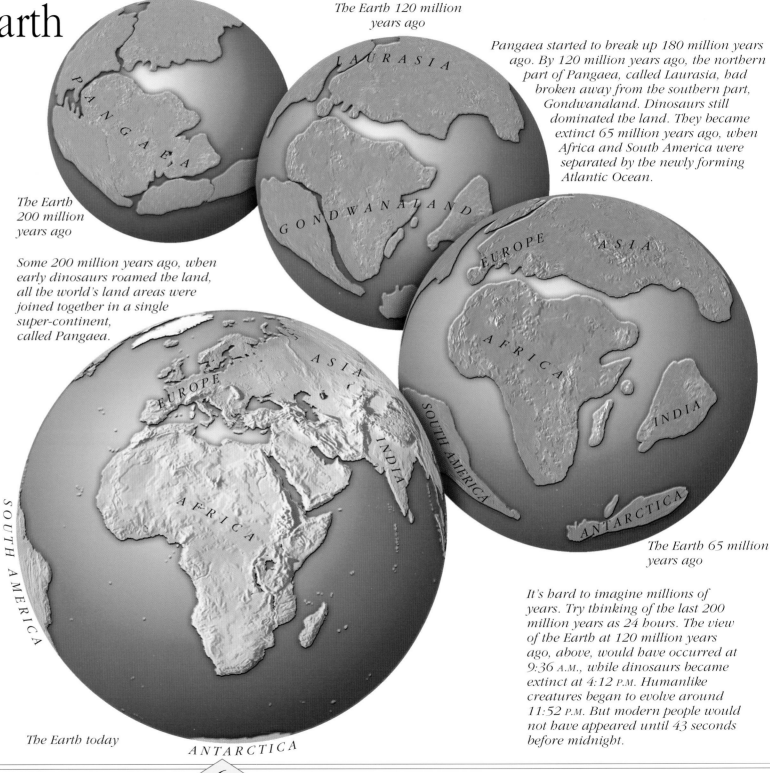

The Earth 120 million years ago

The Earth 200 million years ago

Some 200 million years ago, when early dinosaurs roamed the land, all the world's land areas were joined together in a single super-continent, called Pangaea.

Pangaea started to break up 180 million years ago. By 120 million years ago, the northern part of Pangaea, called Laurasia, had broken away from the southern part, Gondwanaland. Dinosaurs still dominated the land. They became extinct 65 million years ago, when Africa and South America were separated by the newly forming Atlantic Ocean.

The Earth 65 million years ago

The Earth today

It's hard to imagine millions of years. Try thinking of the last 200 million years as 24 hours. The view of the Earth at 120 million years ago, above, would have occurred at 9:36 A.M., while dinosaurs became extinct at 4:12 P.M. Humanlike creatures began to evolve around 11:52 P.M. But modern people would not have appeared until 43 seconds before midnight.

The Restless Earth

The Earth's outer layers, including the crust and the top, rigid layer of the mantle, are split into huge plates. The plates, which are about 62 miles (100 km) thick, rest on the mostly solid mantle. But the mantle also contains some molten material that moves around in slow currents. These currents move the plates and the continents resting on them. Plates move apart along ocean ridges and collide along ocean trenches. Sometimes, when plates push against each other, the rocks between them are squeezed up into mountain ranges. Some plates move alongside each other. They are separated by long faults (cracks) in the Earth's surface.

Plates sometimes move in violent jerks, causing earthquakes. On the average, plates move by only 0.8 to 4 inches (2-10 cm) a year. This sounds slow. But over the course of millions of years, plate movements change the face of the Earth.

Above *The globes show the plates that form the Earth's hard outer layers.*
Below *The Earth contains a solid inner core, made up mainly of iron, and a liquid outer core. The core is about 4,190 miles (6,740 km) across. Around the core is the rocky mantle, which is about 1,800 miles (2,900 km) thick. The crust averages 2.3 miles (6 km) under the oceans and 22 miles (35 km) under the continents.*

Some plates move sideways alongside each other. Most of the time, the jagged plate edges are locked together, but friction occasionally breaks the rocks, causing the plates to move suddenly. This causes earthquakes.

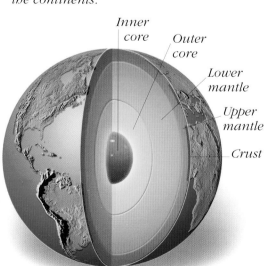

Inner core
Outer core
Lower mantle
Upper mantle
Crust

Sea level | Ocean floor | Ocean ridge | Ocean trench

Magma rising through the crust to form a volcano

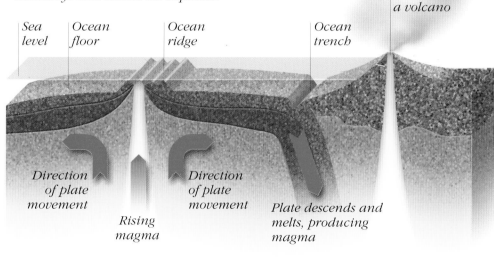

Direction of plate movement
Rising magma
Direction of plate movement
Plate descends and melts, producing magma

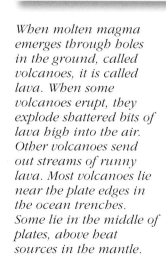

When molten magma emerges through holes in the ground, called volcanoes, it is called lava. When some volcanoes erupt, they explode shattered bits of lava high into the air. Other volcanoes send out streams of runny lava. Most volcanoes lie near the plate edges in the ocean trenches. Some lie in the middle of plates, above heat sources in the mantle.

Hidden beneath the oceans are mountain ranges, called ocean ridges. Deep valleys in the middle of the ridges form the edges of plates that are moving apart. When plates move apart, molten magma from the mantle rises and fills in the gaps. When two plates collide, one plate sinks down beneath the other alongside the ocean trenches, the deepest places in the oceans. The edge of the descending plate melts, and some of the molten rock rises and erupts out of volcanoes on the Earth's surface.

Measuring the Earth

Globes are models of the Earth, showing how our planet looks from a spacecraft. But globes also have names and lines drawn on them. One point marked on the top of a globe is the North Pole, while the point at the bottom is the South Pole. An imaginary line joining the North Pole to the center of the Earth and the South Pole is called the Earth's axis. The axis, around which the Earth rotates, is tilted by 23.5 degrees. Most globes are mounted on stands and tilted, just like Planet Earth.

Halfway between the poles is another imaginary line running around the globe. This line is called the equator. It divides the world into two equal halves, called hemispheres. (The word "hemisphere" means half a sphere.) The equator is a line of latitude, as are the other lines drawn around globes parallel to the equator. Globes also show lines of longitude, which run at right angles to the lines of latitude.

Globes accurately show areas, shapes, directions and distances on Earth. The surfaces of globes also contain networks of lines of latitude and longitude. Every place on Earth has its own latitude and longitude.

The Earth rotates on its axis once every 24 hours. When places on Earth face the Sun, it is day. When they turn away from the Sun, it is night.

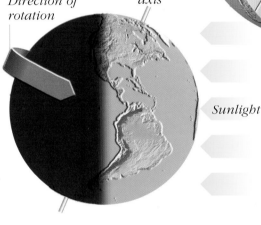

Direction of rotation

The Earth axis

Sunlight

The line of longitude known as prime meridian (0 degrees longitude) runs through Greenwich, England. Time is measured east and west of the prime meridian. The Earth takes 24 hours to rotate once on its axis. So 15 degrees of longitude represents one hour. If you travel east from the prime meridian, 180 degrees represents a gain of 12 hours. But going west, 180 degrees represents a loss of 12 hours. So the world is divided into 24 time zones. The 180 degrees line of longitude is called the International Date Line, marking a time difference of 24 hours. When you cross the line from west to east, you gain a day. When you return from east to west, you lose a day.

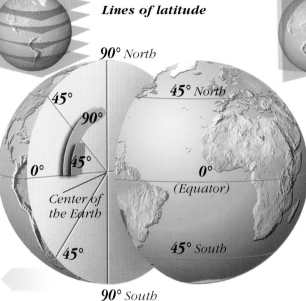

Lines of latitude

90° North

45°

90°

45° North

0°

45°

Center of the Earth

0° (Equator)

45°

45° South

90° South

Latitude and longitude are measured in degrees. The equator is 0 degrees latitude. The North Pole is 90 degrees North, and the South Pole is 90 degrees South. Latitude is measured by the angle made between the equator and the point at the center of the Earth. The cutaway globe, above, shows that latitudes 45° North and South are measured by the 45° angle at the center of the Earth.

Lines of longitude

135° West   180°   135° East

90° West

90° East

45° West

45° East

0°

90°   45°

Center of the Earth

0°

Lines of longitude, or meridians, run at right angles to the lines of latitude, passing through the North and South Poles. They are measured from the center of the Earth, 180 degrees east and 180 degrees west of the prime meridian (0 degrees longitude). Longitudes 45° West and East are shown on the cutaway globe.

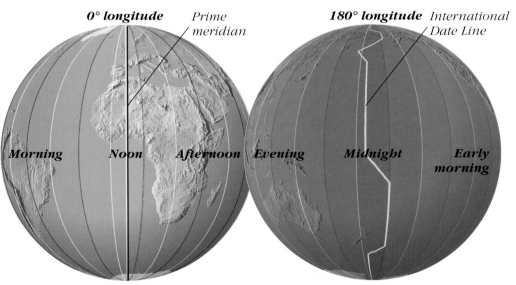

0° longitude   *Prime meridian*

180° longitude   *International Date Line*

Morning   Noon   Afternoon   *Evening   Midnight   Early morning*

Weather and Seasons

June 20 or 21: northern summer and southern winter

March 20 or 21: start of northern spring and southern fall

Sun

The Earth travels around the Sun, completing one journey in 365 days, 5 hours, 48 minutes, and 46 seconds. The diagram shows that when the northern hemisphere tilts toward the Sun, it gets more sunlight. As a result, it is summer. When the southern hemisphere tilts toward the Sun, it is summer in the southern hemisphere and winter in the northern hemisphere.

September 22 or 23: beginning of northern fall and southern spring

December 21 or 22: northern winter and southern summer

The Earth is surrounded by a layer of air, called the atmosphere. Weather is the changing day-to-day state of the atmosphere. For example, a weather forecaster may say that it will be dry in the morning, with rain in the evening. Because the Earth's axis is tilted, regions in the middle latitudes get four seasons every year – spring, summer, fall, and winter. Some places do not have four seasons. They include regions in the low latitudes (near the equator, where it is always hot) or in the high latitudes (near the poles, where it is always cold).

Storms are major weather systems. The most common storms are thunderstorms. Many form near the equator and also where westerly winds meet polar easterlies. Other storms are hurricanes, which form north and south of the equator. These storms can be seen from space.

A huge circular storm, called a hurricane, photographed from a spacecraft.

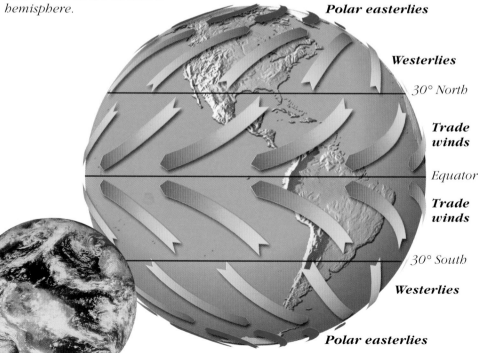

Polar easterlies

Westerlies

30° North

Trade winds

Equator

Trade winds

30° South

Westerlies

Polar easterlies

The photograph of the Earth, far left, shows swirling clouds over our planet. The atmosphere surrounding the Earth is always on the move. The diagram of the world's main wind belts shows that, around the equator, where the Sun's heat is strong, hot air is rising. In the upper atmosphere, this rising air spreads out north and south. The air sinks back to the surface around 30° North and 30° South latitude. At the surface, some of the descending air flows back toward the equator, forming the trade winds. Some flows toward the poles, forming westerly winds. Circular storms called depressions form where the warm westerly winds meet cold easterly winds flowing from the polar regions.

Hot, Cold, Wet, Dry

Climate is the usual, or average, weather of a place. Scientists describe climate according to how hot or cold a place is, and how much rain it gets. There are six main climatic regions.

Mountain climates vary according to the height of the land. This is because the higher you climb up a mountain, the colder it gets. Some mountains have tropical rain climates at the bottom and polar climates at the top.

Places with polar climates have average temperatures of less than 50°F (10°C) in the warmest month. They include areas that are always covered by ice and places with a short, chilly summer. Polar climates have little rain or snow.

Places with cold snowy climates have an average temperature of less than 27°F (-3°C) in the coldest month of the year, but more than 50°F (10°C) in the warmest month. Rain in summer and snow in winter is generally moderate.

Polar climates occur around the poles. Cold snowy climates have long winters. Warm temperate climates have four distinct seasons. Both deserts and dry grasslands have dry climates. Tropical rainy climates occur around the equator. Mountain climates range from tropical to polar, according to height.

Places with warm temperate climates occur in the middle latitudes, and have four marked seasons. Average temperatures do not exceed 64°F (18°C) in the warmest month, but are not less than 27°F (-3°C) in the coldest month.

Places with dry climates may be hot or cold. But they have an average annual precipitation (rain, melted snow, hail, sleet, and so on) of less than 10 inches (250 mm). They include deserts, where years may go by without rain, and dry grasslands.

Places with tropical rainy climates have an average temperature of more than 64°F (18°C) for every month of the year. The rainfall may be heavy and occur in every month of the year. Sometimes the rain may be confined to one rainy season.

The Living World

The cold, treeless areas in polar regions, where plants grow during the short summer, are called tundra.

Mountains contain zones of different plants according to height, including forests, pasture, and ice fields.

Evergreen coniferous forests, with such trees as fir, pine, and spruce, grow in places with cold, snowy climates.

Temperate forests, with trees that shed their leaves in the fall, grow in places with warm temperate climates.

Climate largely determines the plants and animals in any region. For example, the cold, frozen tundra is treeless, and only animals that can survive the cold live in polar regions.

Temperate, mainly treeless grasslands occur in places with dry climates. They include prairie and steppe.

Tropical rainforests grow in regions with tropical climates, with plenty of rainfall, often throughout the year.

Savanna, tropical grassland with scattered trees, occurs in tropical rainy regions with a distinct dry season.

Deserts have little rainfall. No rain may fall for several years, but then one cloudburst may flood large areas.

People have changed some regions. For example, most of the warm temperate forests have been cut down to make space for farming. Today, tropical rainforests are also being cut down. These forests contain more than half of the world's living things. Scientists see their destruction as a major disaster.

Population

In 1999, the world's population passed the 6 billion mark. Large areas are too dry or too cold to support more than a few people, while in other areas people are crowded together. Only about 10 percent of the land on Earth is suitable for growing crops, and another 20 percent is used for grazing farm animals.

Areas with good soil and a pleasant climate contain large numbers of people, especially in cities. For example, about two-thirds of the land in Australia is almost empty of people, because it is too dry. More than 60 percent of Australians live in five cities: Sydney, Melbourne, Brisbane, Perth, and Adelaide. From space, cities show up as sparkling dots of light at night.

Hong Kong is a thriving island city off the coast of South China, with a population of over 4.5 million people. With only limited land available for the growing population, most people live in high rise buildings. Many Asian cities have populations of over 1 million. Most of the world's most rapidly growing cities are in this region.

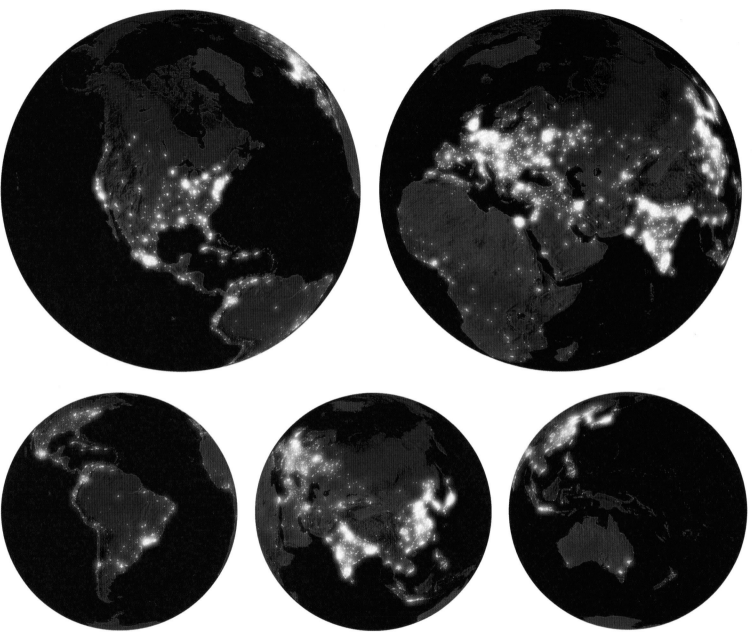

Imagine looking down from space on Earth at night. If everyone on Earth stood out in the open holding a lighted candle, thickly populated areas of the world would appear as bright points of light. Thinly populated areas would look dark and empty.

The darkened globes shown above reveal brightly lit areas where people are crowded together. You can see how the northeastern United States and west-central Europe are thickly populated. But most of northern Africa, which is desert, looks empty of people.

Asia contains some of the world's most thickly populated areas, especially in India, Bangladesh, eastern China, and Southeast Asia. China and India have more people than any other countries. In contrast, most of South America and Australia are thinly populated.

The shortest distance between two points is a straight line. But straight lines drawn on many world maps are not the shortest distances. This is because the Earth is round and not flat. To find the shortest distance between places on Earth, take a piece of string and stretch it across a globe between any two places. The line followed by the string is called a great-circle route. (A great circle divides the Earth into two equal halves.) The globes at right and below show some great-circle routes between cities.

To London
To London
To New York
To Rio de Janeiro
To Los Angeles
Tokyo
Cape Town
Sydney
To Los Angeles
To Rio de Janeiro

To Tokyo
To Tokyo
To Tokyo
To Sydney
To Sydney
Los Angeles
London
New York City
Rio de Janeiro
Cape Town

Modern canals save ships long sea journeys. **Right** the globe shows that before the Panama Canal was completed in 1914, the sea journey from New York City to San Francisco around South America was more than 13,000 miles (20,900 km). The canal reduced the journey to less than 5,200 miles (about 8,300 km). **Center right** the Suez Canal, Egypt, connects the Mediterranean and Red Seas. It saves oil tankers from the Gulf from having to sail around Africa to reach western Europe. **Far right** Greece's Corinth Canal is a shortcut for ships sailing from the Aegean to the Ionian Sea.

Los Angeles
New York City
Panama Canal

Rotterdam
The Gulf
Suez Canal

Getting Around

A round 200 years ago, people seldom went far from their homes. Most of them lived by farming, and the longest journey many of them took was a walk to the nearest town on market day. But the invention of engine-powered machines starting in the late 18th century caused a rapid revolution in transportation. Today, heavy raw materials for industry are shipped around the world by road and sea, while people fly in jet airliners to countries on the other side of the globe. Some experts predict that, before long, tourists will be taking trips in space!

Transportation takes three main forms – land, sea, and air. Land transportation includes cars, trains, and huge trucks that carry heavy goods. Water transportation includes the movement of bulky goods in cargo ships, such as the huge tankers that carry oil to industrial countries. Air transportation is efficient for moving expensive, lightweight, and perishable goods as well as passengers. Faster, cheaper transportation has helped increase international trade, which has raised the living standards of developing countries.

Canada and Greenland

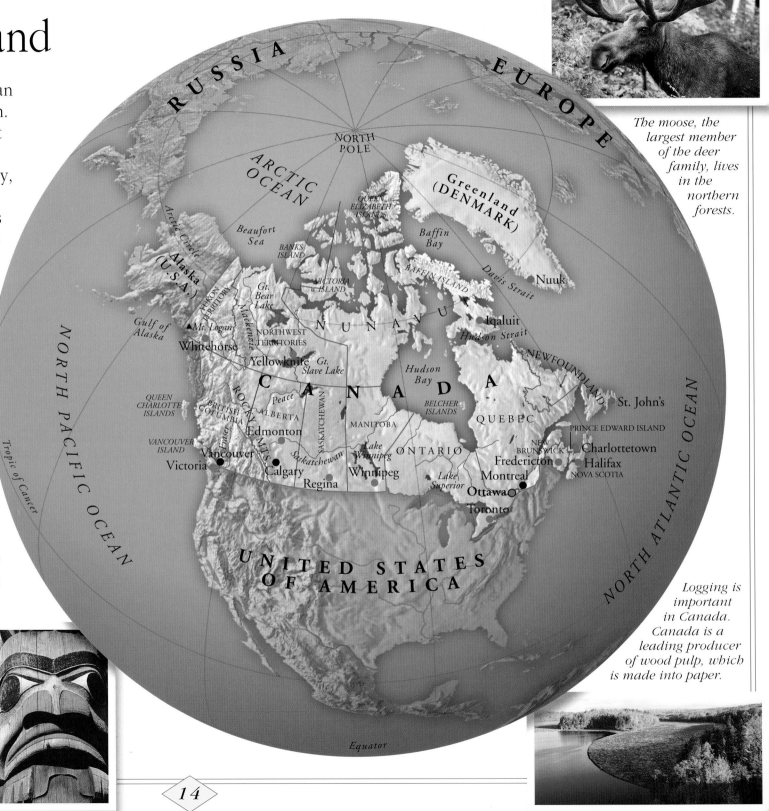

Let's take a trip to Canada. From space, we can see that Canada borders the icy Arctic Ocean. East of Canada is Greenland, the world's largest island. Most of Greenland is buried by ice.

Canada is the world's second largest country, after Russia. But it has only about 30 million people. The United States has nearly nine times as many people as Canada. The far north is too cold even for trees. But vast forests of evergreen trees cover central Canada and the western mountains. Animals such as bears, beavers, moose, and wolves live in the northern forests. Mount Logan, Canada's highest peak, lies near the border with Alaska. Most Canadians live within about 200 miles (320 km) of the United States border. This region contains most of Canada's fertile farmland and its largest cities, including the capital, Ottawa.

The moose, the largest member of the deer family, lives in the northern forests.

Above *Wheat is the chief crop on the prairies east of the Rocky Mountains.*

Below *The Inuit in northern Canada use sleds pulled by Huskies.*

A Native American totem pole from western Canada.

Logging is important in Canada. Canada is a leading producer of wood pulp, which is made into paper.

United States of America

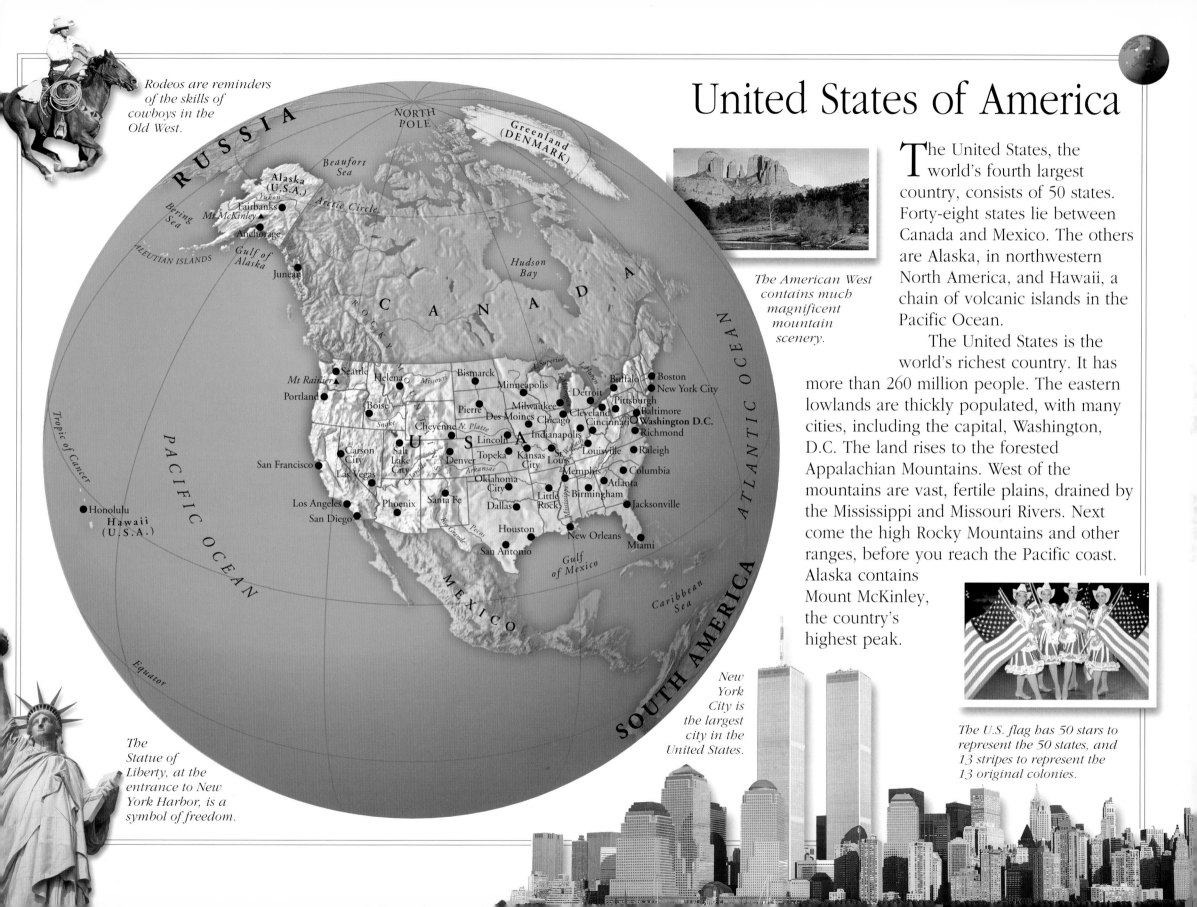

Rodeos are reminders of the skills of cowboys in the Old West.

RUSSIA

NORTH POLE

Greenland (DENMARK)

Beaufort Sea

Alaska (U.S.A.)

Arctic Circle

Bering Sea

Yukon

Fairbanks

Mt McKinley ▲

Anchorage

ALEUTIAN ISLANDS

Gulf of Alaska

Juneau

CANADA

Hudson Bay

ROCKY MOUNTAINS

L.Superior

L.Huron

Mt Rainier ▲ Seattle

Helena

Missouri

Bismarck

Minneapolis

Buffalo

Boston

New York City

Portland

Boise

Pierre

Milwaukee

Detroit

Pittsburgh

Baltimore

Cleveland

Cincinnati

Washington D.C.

Snake

Cheyenne N. Platte

Des Moines

Chicago

Richmond

U S A

Lincoln

Indianapolis

Wabash

Carson City

Salt Lake City

Denver

Topeka

Kansas City

St. Louis

Louisville

Raleigh

San Francisco

Arkansas

Las Vegas

Colorado

Oklahoma City

Memphis

Columbia

Los Angeles

Phoenix

Santa Fe

Dallas

Little Rock

Atlanta

Birmingham

Jacksonville

San Diego

Rio Grande

Pecos

Houston

Mississippi

New Orleans

Miami

Honolulu

Hawaii (U.S.A.)

San Antonio

Gulf of Mexico

PACIFIC OCEAN

Tropic of Cancer

MEXICO

Caribbean Sea

ATLANTIC OCEAN

SOUTH AMERICA

Equator

The American West contains much magnificent mountain scenery.

The United States, the world's fourth largest country, consists of 50 states. Forty-eight states lie between Canada and Mexico. The others are Alaska, in northwestern North America, and Hawaii, a chain of volcanic islands in the Pacific Ocean.

The United States is the world's richest country. It has more than 260 million people. The eastern lowlands are thickly populated, with many cities, including the capital, Washington, D.C. The land rises to the forested Appalachian Mountains. West of the mountains are vast, fertile plains, drained by the Mississippi and Missouri Rivers. Next come the high Rocky Mountains and other ranges, before you reach the Pacific coast. Alaska contains Mount McKinley, the country's highest peak.

The Statue of Liberty, at the entrance to New York Harbor, is a symbol of freedom.

New York City is the largest city in the United States.

The U.S. flag has 50 stars to represent the 50 states, and 13 stripes to represent the 13 original colonies.

Southern North America

North America is the third largest continent after Asia and Africa. It contains not only Canada and the United States, but also Mexico, the seven countries of Central America, and the island countries in the Caribbean Sea.

Mexico is the largest country in southern North America. If you take a look at Mexico from space, you will see that it contains mountain ranges, a central plateau dotted with smoking volcanoes, and some coastal lowlands. South of Mexico, the countries of Central America form a bridge, linking North and South America. The Panama Canal cuts through this bridge, providing a shortcut for ships sailing from the Atlantic to the Pacific Ocean.

The Caribbean islands have beautiful beaches that attract many tourists. But danger sometimes lurks in the Caribbean. Great hurricanes often batter the coasts, and volcanoes occasionally erupt, driving people from their homes.

Above Many tourists visit island resorts in the Caribbean Sea.
Below Most Mexicans have both white and Native American ancestors.

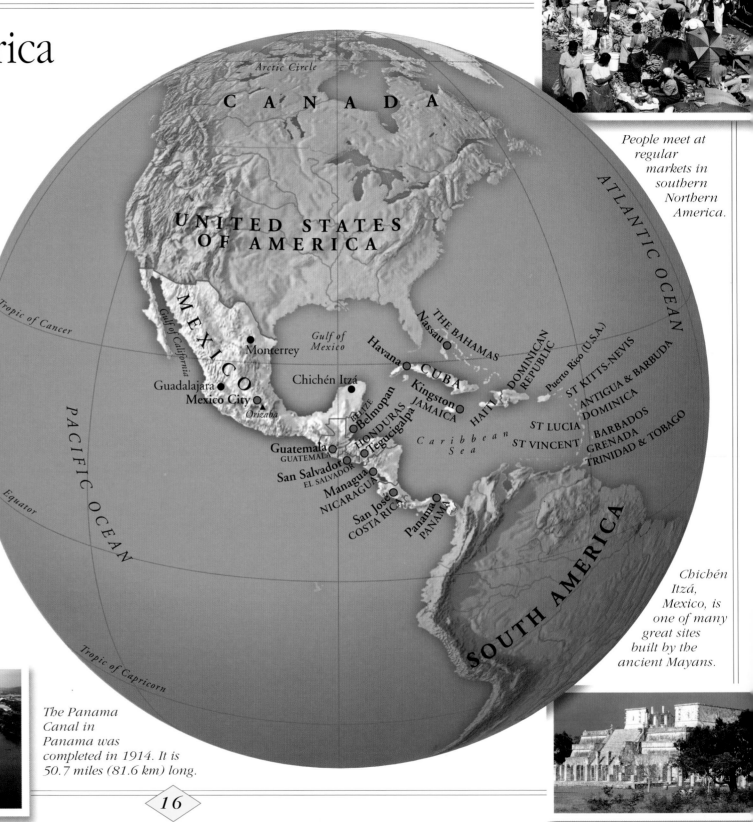

People meet at regular markets in southern Northern America.

Chichén Itzá, Mexico, is one of many great sites built by the ancient Mayans.

The Panama Canal in Panama was completed in 1914. It is 50.7 miles (81.6 km) long.

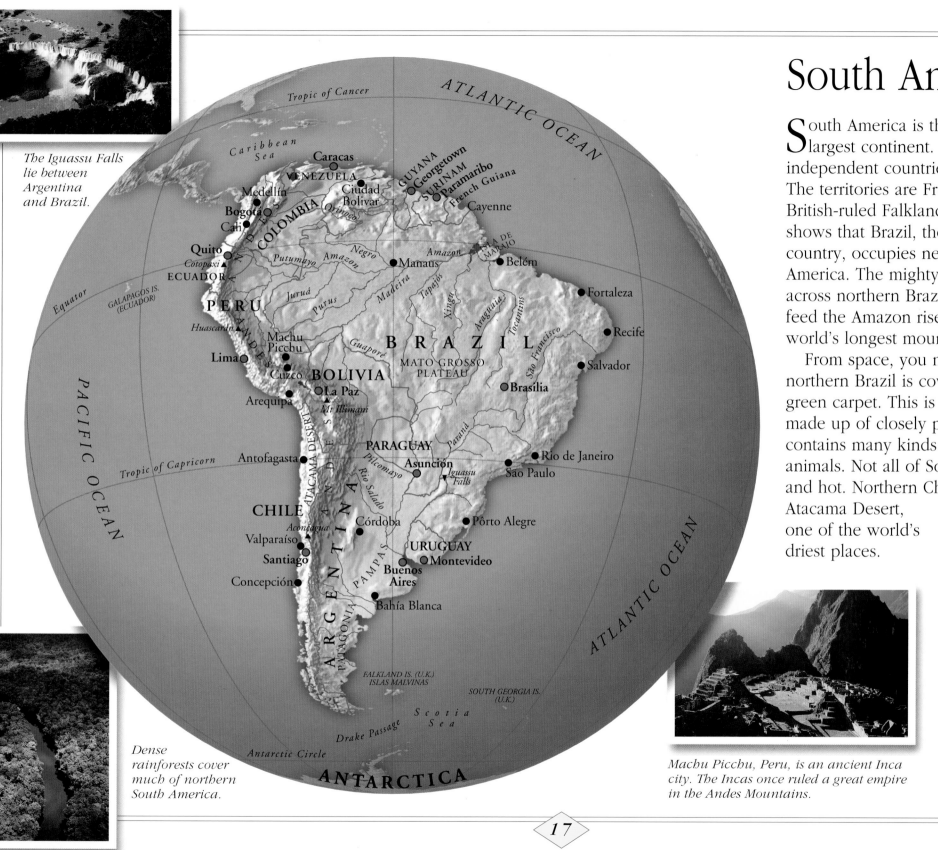

The Iguassu Falls lie between Argentina and Brazil.

Dense rainforests cover much of northern South America.

Machu Picchu, Peru, is an ancient Inca city. The Incas once ruled a great empire in the Andes Mountains.

South America

South America is the world's fourth largest continent. It consists of 12 independent countries and two territories. The territories are French Guiana and the British-ruled Falkland Islands. The globe shows that Brazil, the world's fifth largest country, occupies nearly half of South America. The mighty Amazon River flows across northern Brazil. Many rivers that feed the Amazon rise in the Andes, the world's longest mountain range.

From space, you might think that northern Brazil is covered by a huge green carpet. This is a vast rainforest made up of closely packed trees. It contains many kinds of plants and animals. Not all of South America is wet and hot. Northern Chile contains the Atacama Desert, one of the world's driest places.

Many people in South America are descendants of Native Americans and Europeans.

Map labels

ATLANTIC OCEAN
PACIFIC OCEAN
Tropic of Cancer
Caribbean Sea
Equator
Tropic of Capricorn
Antarctic Circle
ANTARCTICA
Scotia Sea
Drake Passage

VENEZUELA — Caracas
Medellín
Ciudad Bolívar
GUYANA — Georgetown
SURINAM — Paramaribo
French Guiana — Cayenne
COLOMBIA — Bogotá, Cali
Orinoco
Quito
ECUADOR
Cotopaxi
GALAPAGOS IS. (ECUADOR)
PERU
Huascarán
Putumayo
Amazon
Negro
Juruá
Purus
Madeira
Tapajós
Xingu
Araguaia
Tocantins
Manaus
ILHA DE MARAJO
Belém
Fortaleza
Recife
Salvador
São Francisco
Machu Picchu
Lima
Cuzco
Guaporé
BRAZIL
MATO GROSSO PLATEAU
Brasília
BOLIVIA
La Paz
Mt Illimani
Arequipa
ATACAMA DESERT
Paraná
PARAGUAY
Asunción
Pilcomayo
Rio Salado
Iguassu Falls
Rio de Janeiro
São Paulo
Antofagasta
CHILE
Aconcagua
Córdoba
ANDES
ARGENTINA
Pôrto Alegre
URUGUAY
Montevideo
Valparaíso
Santiago
Buenos Aires
Concepción
PAMPAS
PATAGONIA
Bahía Blanca
FALKLAND IS. (U.K.) ISLAS MALVINAS
SOUTH GEORGIA IS. (U.K.)

Europe

Europe is the sixth largest continent, but it is densely populated with many cities. It contains 42 complete countries, containing about 580 million people. It also includes one-fourth of Turkey and European Russia. European Russia has a population of about 120 million.

Northern Europe borders the icy Arctic Ocean. Evergreen forests in the north are the home of such animals as brown bears, reindeer, and wolves. The forests of central and southern Europe have been largely cut down to make way for farmland and cities. The chief mountain range in western Europe is the Alps. South of the Alps lie Europe's Mediterranean countries, which have hot, dry summers.

Tulips are famous products of the Netherlands.

The Matterhorn is one of the magnificent peaks in the Alps.

Europe has several small countries that are hard to find on the map. On this globe, 14 small countries are shown by numbers. Can you find the 14 countries listed below on the globe?

1 Andorra	*7 Luxembourg*
2 Bosnia &	*8 Macedonia*
Herzegovina	*9 Moldova*
3 Croatia	*10 Monaco*
4 Gibraltar	*11 San Marino*
(U.K.)	*12 Slovenia*
5 Kaliningrad	*13 Vatican City*
(Russia)	*14 Yugoslavia*
6 Liechtenstein	

Stonehenge in southern England is one of Europe's best known ancient sites. Europe has a long history, with many monuments dating back 2,000 years and more.

A church in Bavaria, Germany, is a reminder that Christianity is the chief religion of Europe.

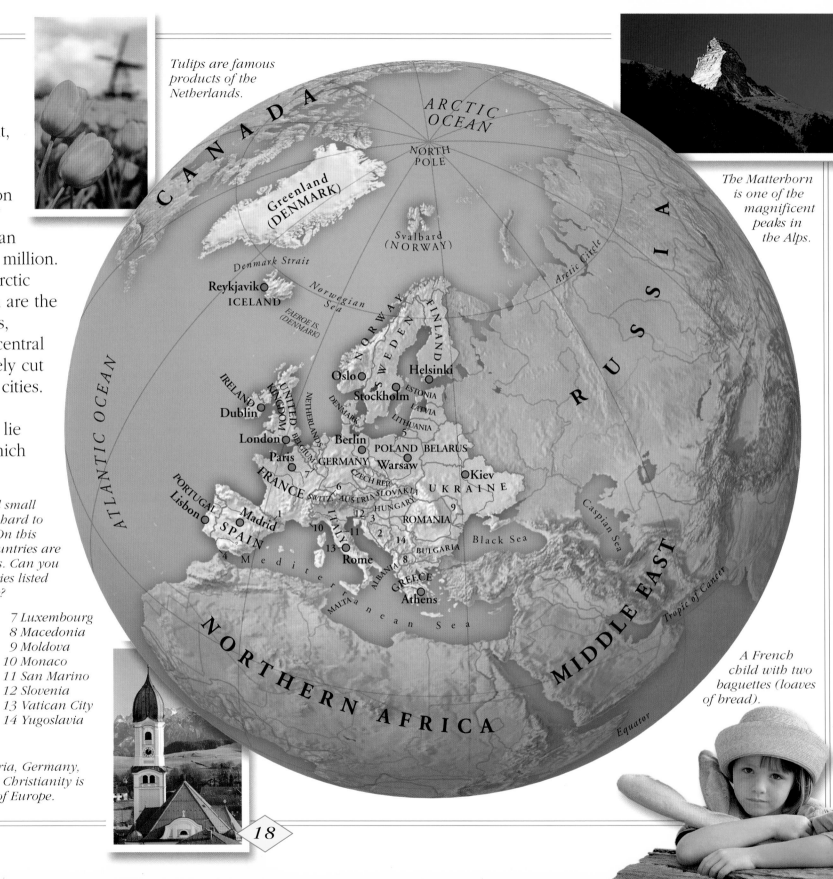

ARCTIC OCEAN

CANADA

NORTH POLE

Greenland (DENMARK)

Svalbard (NORWAY)

Denmark Strait

Reykjavik
ICELAND

FAEROE IS. (DENMARK)

Norwegian Sea

Arctic Circle

RUSSIA

NORWAY

SWEDEN

FINLAND

Oslo

Helsinki

Stockholm

ESTONIA

LATVIA

LITHUANIA

ATLANTIC OCEAN

IRELAND

UNITED KINGDOM

Dublin

NETHERLANDS

DENMARK

Berlin

POLAND

BELARUS

London

BELGIUM

GERMANY

Warsaw

Paris

7

Kiev

FRANCE

SWITZ.

CZECH REP.

AUSTRIA

SLOVAKIA

UKRAINE

6

HUNGARY

9

PORTUGAL

12

3

ROMANIA

Caspian Sea

Lisbon

Madrid

1

2

14

SPAIN

ITALY

10

11

Black Sea

13

Rome

BULGARIA

4

8

Mediterranean Sea

MALTA

ALBANIA

GREECE

Athens

MIDDLE EAST

NORTHERN AFRICA

Tropic of Cancer

Equator

A French child with two baguettes (loaves of bread).

Russia and Its Neighbors

Wooden buildings are common in northern Russia.

CANADA

Alaska (U.S.A.)

Greenland (DENMARK)

NORTH POLE

ATLANTIC OCEAN

ARCTIC OCEAN

Arctic Circle

NEW SIBERIAN ISLANDS

FRANZ JOSEPH LAND

SEVERNAYA ZEMLYA

Laptev Sea

Kolyma

CHUCHI PENINSULA

Bering Sea

Petropavlovsk-Kamchatskiy

Barents Sea

NOVAYA ZEMLYA

Kara Sea

Lena

Yakutsk

Sea of Okhotsk

KURIL ISLANDS

SAKHALIN

Murmansk

CENTRAL SIBERIAN PLATEAU

Vilyuy

Aldan

RUSSIA

St Petersburg

URAL MOUNTAINS

Lena

Vladivostok

Minsk
BELARUS

Moscow

Ob'

Yenisey

Angara

L.Baikal

Ulan Ude

PACIFIC OCEAN

Kiev
UKRAINE

Nizhniy Novgorod

Kazan

Yekaterinburg

Irtysh

Ob'

Irkutsk

MOLDOVA
Chisinau

Volga

Omsk

Novosibirsk

Volgograd

Astana

Black Sea

KAZAKSTAN

L.Balkhash

Tropic of Cancer

GEORGIA
Tbilisi

ARMENIA
Yerevan

Caspian Sea

Aral Sea

UZBEKISTAN

Tashkent

EASTERN ASIA

Sea of Japan

AZERBAIJAN
Baku

TURKMENISTAN

Bishkek
KYRGYZSTAN

Ashgabat

TAJIKISTAN
Dushanbe

SOUTHERN ASIA

Equator

Russia's long, freezing winters make warm clothing a necessity.

The world map is always changing. In December 1991, one country called the Soviet Union broke up into 15 countries. One of them is Russia, the world's largest country. European Russia lies to the west of the Ural Mountains. It contains about four-fifths of the country's people. Asian Russia, or Siberia, which makes up three-fourths of the country, is thinly populated.

Of the other countries created from the Soviet Union, Belarus, Estonia, Latvia, Lithuania, Moldova, and Ukraine are in Europe. The other eight countries – Armenia, Azerbaijan, Kazakstan, Kyrgyzstan, Georgia, Tajikistan, Turkmenistan, and Uzbekistan – lie mainly or entirely in the continent of Asia.

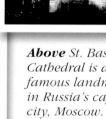

Above St. Basil's Cathedral is a famous landmark in Russia's capital city, Moscow.
Left Northern Russia is a land of forests and lakes.

Left The Cosmos space monument commemorates Russia's achievements in space exploration.

Northern Africa

Imagine flying across the middle of North Africa from the Atlantic Ocean to the Red Sea. Your journey would take you across the Sahara, the world's largest desert. You would see very few signs of life until you reached the well-cultivated Nile valley in Egypt and Sudan. The Nile is the world's longest river. Its water comes from the rainy highlands to the south.

Some areas in the north, where the Mediterranean Sea separates Africa from Europe, have enough rain for farming. Northern North Africa is the land of the Arabs, who believe in Islam. South of the Sahara, the land merges into grassland with scattered trees, called savanna. The far south is hot and rainy, but most of the former forests have been cut down by farmers.

Camels are used as beasts of burden throughout northern Africa.

Date palms are common plants in the dry countries of North Africa.

Ancient Egypt grew up in the Nile valley around 5,000 years ago. Farmers used river water to grow crops. Great ancient Egyptian monuments, including the pyramids, above, and the sphinx, right, can be seen near Cairo.

Many soldiers in Morocco are skilled horsemen.

Southern Africa

Southern Africa contains 24 of Africa's 53 independent countries, including Madagascar, the world's fourth largest island. A trip from north to south would take you from hot, rainy regions around the equator to places with a mild climate on the southern tip of Africa.

Your trip would take you through rainforests, with screeching birds and monkeys. To the south are vast areas of savanna, the home of such animals as antelopes, elephants, giraffes, leopards, lions, rhinoceroses, and zebras. Southwestern Africa contains the bleak Namib Desert and also the Kalahari, a harsh, arid area, mainly in Botswana. In the far south is South Africa. South Africa has many mines, factories, farms, and large cities.

Lions are the leading animal hunters in the savanna regions.

Zebras and elephants live in southern Africa.

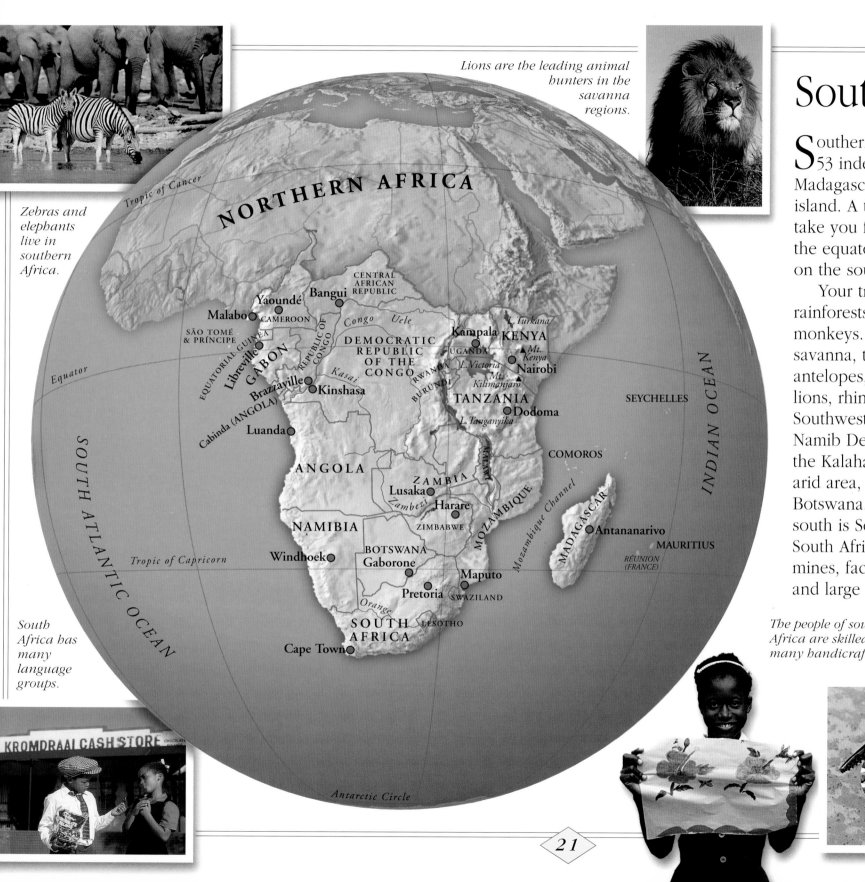

South Africa has many language groups.

The people of southern Africa are skilled at many handicrafts.

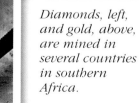

Diamonds, left, and gold, above, are mined in several countries in southern Africa.

Middle East

Asia, the world's largest continent, is divided into four main regions. One of them is called the Middle East, or southwestern Asia. A view from space shows that most of the Middle East consists of empty desert or parched grassland, where water is scarce. Farmland occurs in the northern uplands and along rivers, especially the Tigris and Euphrates in Turkey and Iraq. Other farms are irrigated, or artificially watered. Sometimes, the water comes from deep wells. In other areas, water is piped long distances to dry lands. Water is not the only thing hidden beneath the ground. The rocks in some areas contain oil and natural gas. Oil and gas sales have brought great wealth to several Middle Eastern countries.

The Middle East is the home of three great religions: Judaism, the chief religion of Israel; Christianity; and Islam. More than four out of every five people in the Middle East are Muslims (followers of Islam).

Camels can go for several days without water.

Jerusalem is a holy city for Jews, Christians, and Muslims.

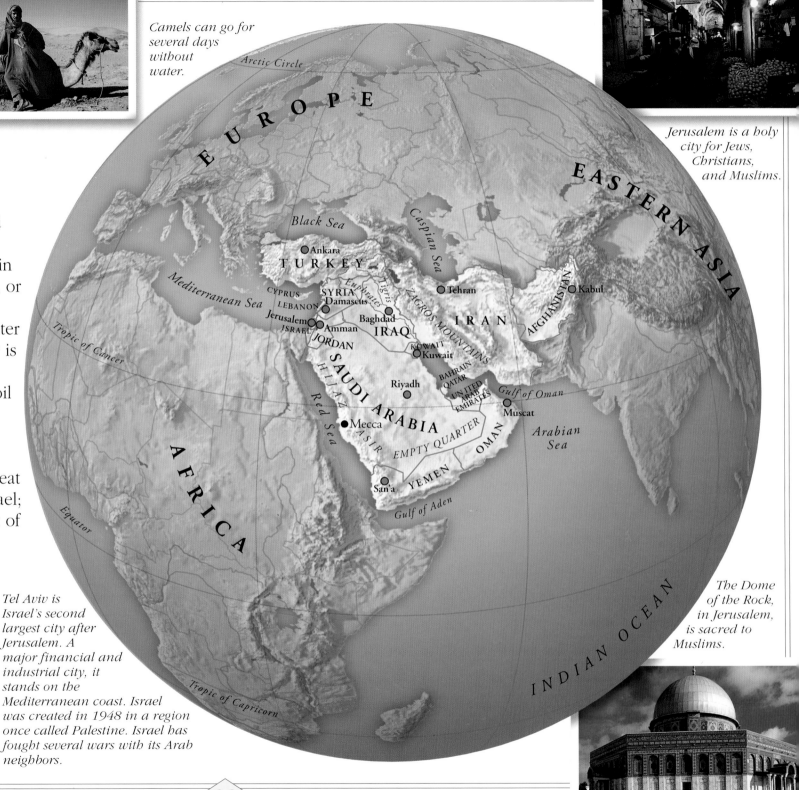

The Dome of the Rock, in Jerusalem, is sacred to Muslims.

Tel Aviv is Israel's second largest city after Jerusalem. A major financial and industrial city, it stands on the Mediterranean coast. Israel was created in 1948 in a region once called Palestine. Israel has fought several wars with its Arab neighbors.

Southern Asia

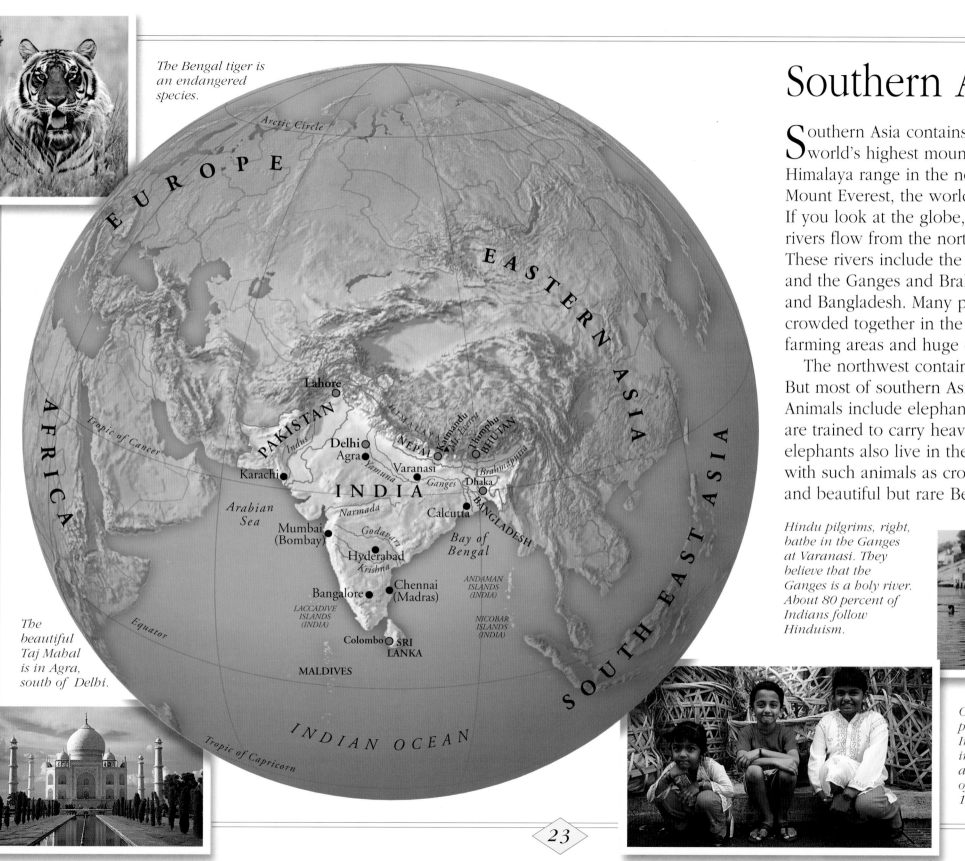

The Bengal tiger is an endangered species.

EUROPE

AFRICA

Arctic Circle

EASTERN ASIA

Tropic of Cancer

Lahore

PAKISTAN

Indus

HIMALAYA
Kathmandu
NEPAL
Mt. Everest
Thimphu
BHUTAN

Delhi
Agra
Yamuna
Varanasi
Ganges
Brahmaputra
Dhaka

Karachi

INDIA

BANGLADESH

Arabian Sea

Narmada

Calcutta

Mumbai
(Bombay)

Godavari

Bay of Bengal

Hyderabad

Krishna

ANDAMAN ISLANDS (INDIA)

Bangalore

Chennai
(Madras)

LACCADIVE ISLANDS (INDIA)

NICOBAR ISLANDS (INDIA)

Equator

Colombo
SRI LANKA

MALDIVES

SOUTH EAST ASIA

Tropic of Capricorn

INDIAN OCEAN

The beautiful Taj Mahal is in Agra, south of Delhi.

Southern Asia contains many of the world's highest mountains. The Himalaya range in the north contains Mount Everest, the world's highest peak. If you look at the globe, you will see that rivers flow from the northern mountains. These rivers include the Indus in Pakistan and the Ganges and Brahmaputra in India and Bangladesh. Many people live crowded together in the river valleys, in farming areas and huge cities.

The northwest contains some deserts. But most of southern Asia is hot and wet. Animals include elephants, many of which are trained to carry heavy loads. Some elephants also live in the wild, together with such animals as crocodiles, snakes, and beautiful but rare Bengal tigers.

Hindu pilgrims, right, bathe in the Ganges at Varanasi. They believe that the Ganges is a holy river. About 80 percent of Indians follow Hinduism.

Only China has more people than India. India's population is increasing quickly, and more than a third of its people are under 15 years of age.

Eastern Asia

Eastern Asia includes China, the world's third largest country. China is the only country with more than a billion people. Eastern Asia also includes Japan, Asia's richest country.

On the globe, you will see that western China has many mountain ranges. It also includes the windswept plateau (tableland) of Tibet, and several deserts, including the cold Gobi Desert in northern China and Mongolia. Most Chinese people live in the east, especially in the valleys of such rivers as the Huang He, the Chang Jiang, and the Xi Jiang. Eastern Asia contains many great cities, including Tokyo, capital of Japan, together with Shanghai, on the east coast of China, and Beijing, the Chinese capital.

Above *Japanese geishas are entertainers. They are trained in the arts of dance, music, and conversation.*

Above right *The former British territory of Hong Kong is a port and a leading financial and industrial center. Britain returned Hong Kong to China in 1997.*

The Great Wall of China.

Japan has high-speed, bullet-shaped electric trains.

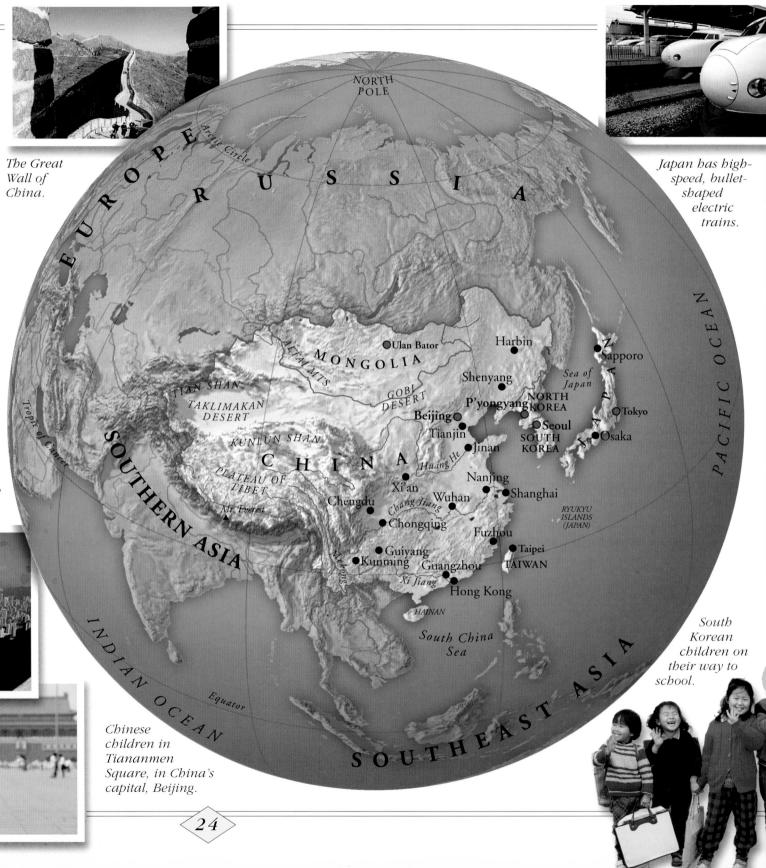

Chinese children in Tiananmen Square, in China's capital, Beijing.

South Korean children on their way to school.

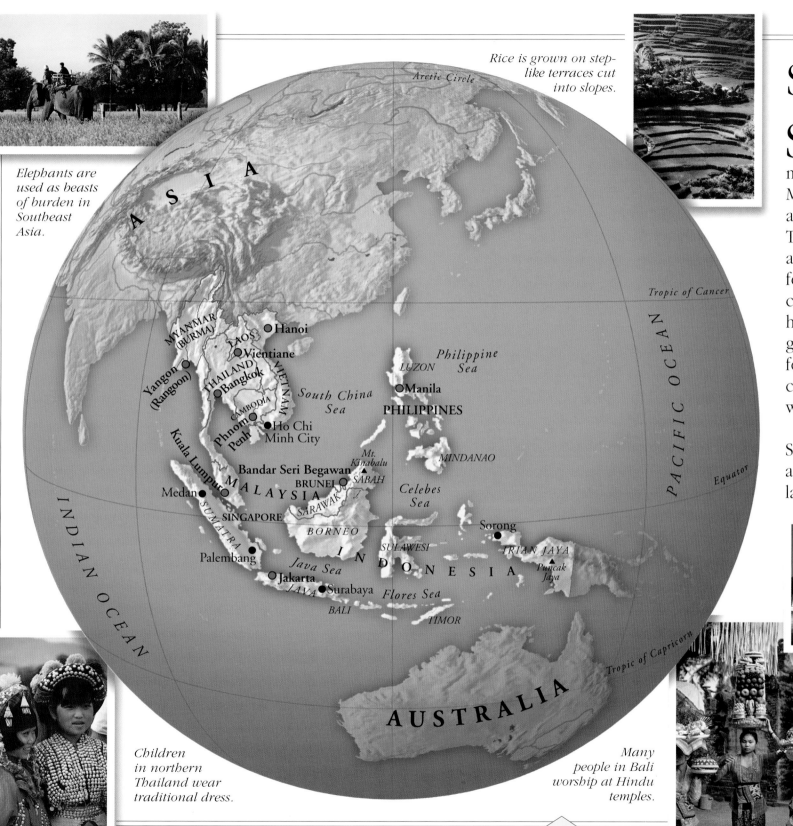

Elephants are used as beasts of burden in Southeast Asia.

Rice is grown on step-like terraces cut into slopes.

Children in northern Thailand wear traditional dress.

Many people in Bali worship at Hindu temples.

Southeast Asia

Southeast Asia consists of some countries on the southern Asian mainland, together with many islands. Much of the land is mountainous, with active volcanoes on some of the islands. The climate is hot and rainy, and forests and swamps cover large areas. But many forests have been cut down or burned to create farms. The number of wild animals has decreased, although monkeys, gibbons, and orangutans still live in the forests. One animal found in Indonesia is called the Komodo dragon. It is the world's largest lizard.

Farming is the main activity in Southeast Asia, and rice is the chief crop and staple food. But the region also has large cities, with many factories.

Left _A Buddhist temple in Myanmar (Burma)._

Below _Singapore is a busy center of industry and trade._

Australia and New Zealand

Australia, the smallest of the world's seven continents, and the island country of New Zealand lie in the southern hemisphere. Australia is the flattest continent. If you fly from east to west across Australia, you will see that the Great Dividing Range separates the east coast from the rest of the country. The Great Dividing Range is Australia's main upland area. It contains Australia's highest peak, Mount Kosciuszko. Beyond the Great Dividing Range lie the flat central lowlands, where farmers raise cattle and sheep. To the west is a huge plateau, or flat tableland, broken by some low mountains. The west is mainly desert, and few people live there.

New Zealand has more varied scenery. North Island has active volcanoes. South Island contains a mountain range called the Southern Alps. Its highest peak is Mount Cook.

Koalas live in trees in Australia.

Lamb and wool are major exports from New Zealand.

Above *Maoris were the original people of New Zealand.*

Right *Uluru (Ayers Rock), in central Australia, is regarded as a sacred place by the Aboriginal people of Australia.*

Left *Sydney is Australia's largest city. Its famous Opera House consists of white concrete shells that look like huge sails.*

Kangaroos are now protected in Australia.

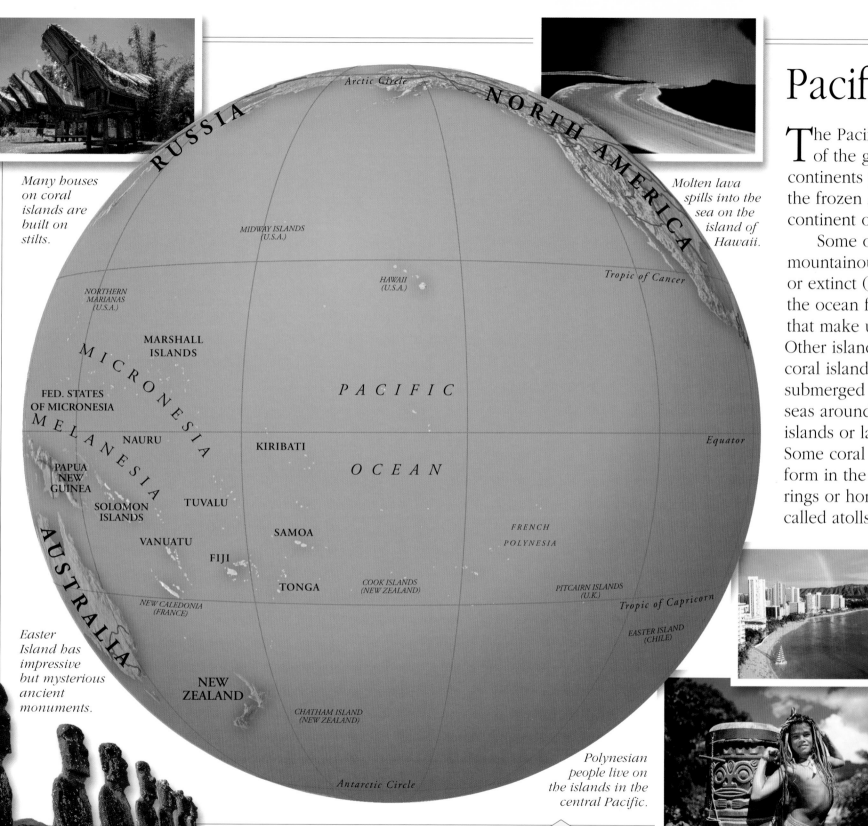

Many houses on coral islands are built on stilts.

Molten lava spills into the sea on the island of Hawaii.

RUSSIA

NORTH AMERICA

Arctic Circle

MIDWAY ISLANDS (U.S.A.)

Tropic of Cancer

HAWAII (U.S.A.)

NORTHERN MARIANAS (U.S.A.)

MARSHALL ISLANDS

MICRONESIA

PACIFIC

FED. STATES OF MICRONESIA

MELANESIA

NAURU

KIRIBATI

Equator

PAPUA NEW GUINEA

OCEAN

SOLOMON ISLANDS

TUVALU

VANUATU

SAMOA

FRENCH POLYNESIA

FIJI

AUSTRALIA

TONGA

COOK ISLANDS (NEW ZEALAND)

PITCAIRN ISLANDS (U.K.)

NEW CALEDONIA (FRANCE)

Tropic of Capricorn

EASTER ISLAND (CHILE)

NEW ZEALAND

CHATHAM ISLAND (NEW ZEALAND)

Easter Island has impressive but mysterious ancient monuments.

Antarctic Circle

Polynesian people live on the islands in the central Pacific.

Pacific Ocean

The Pacific Ocean covers about a third of the globe. It is bigger than all the continents put together. It stretches from the frozen Arctic Ocean to the icy continent of Antarctica.

Some of the Pacific Ocean islands are mountainous. They are the tops of active or extinct (dead) volcanoes that rise from the ocean floor. For example, the islands that make up Hawaii are all volcanic. Other islands are low-lying. They are coral islands which form on the tops of submerged volcanoes or in the shallow seas around other islands or land masses. Some coral islands form in the shapes of rings or horseshoes, called atolls.

Above *Tiny jellylike creatures called coral polyps build hard outer skeletons. Layers of these structures form coral islands, below.*

Left *Waikiki Beach, on Oahu Island, Hawaii, is a major seaside resort.*

Atlantic Ocean

The Atlantic is the world's second largest ocean. It is a busy ocean, with ships crisscrossing its waters, carrying goods from one continent to another. It also contains large fishing grounds, although overfishing has caused fish stocks to drop to low levels.

The ocean's chief feature is mainly hidden from view. This is the mid-Atlantic ridge, a huge mountain range rising from the deep ocean floor and running north-south through the ocean. The ridge is the place where plates are moving apart and where molten lava is rising to the surface to form new crustal rock. As a result, the Atlantic Ocean is becoming wider by about one inch (2.5 cm) a year. The volcanic island of Iceland rises from the ridge. It, too, is becoming wider as the plates on either side move apart. Newfoundland, in eastern Canada, the British Isles, and the sunny islands of the Caribbean Sea are also Atlantic islands.

Cape Verde is a country in the North Atlantic Ocean. It is located about 400 miles (640 km) west of Dakar, the capital of Senegal. It consists of ten large volcanic islands and five islets. The people are descendants of African slaves or of mixed African and Portuguese descent. The islands have a dry climate and water is scarce. Most people are poor and, when long droughts occur, many people starve.

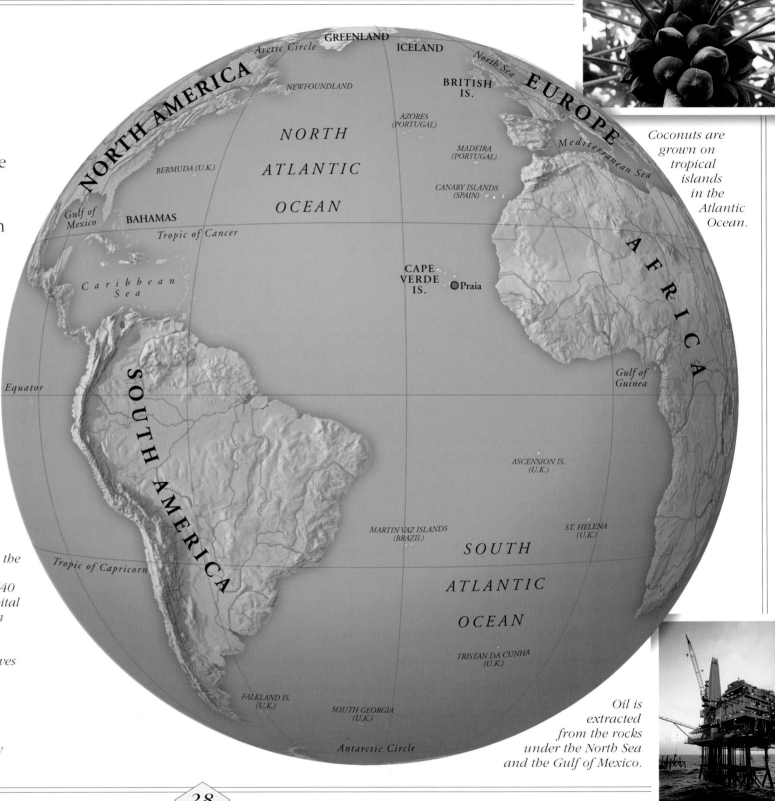

Coconuts are grown on tropical islands in the Atlantic Ocean.

Oil is extracted from the rocks under the North Sea and the Gulf of Mexico.

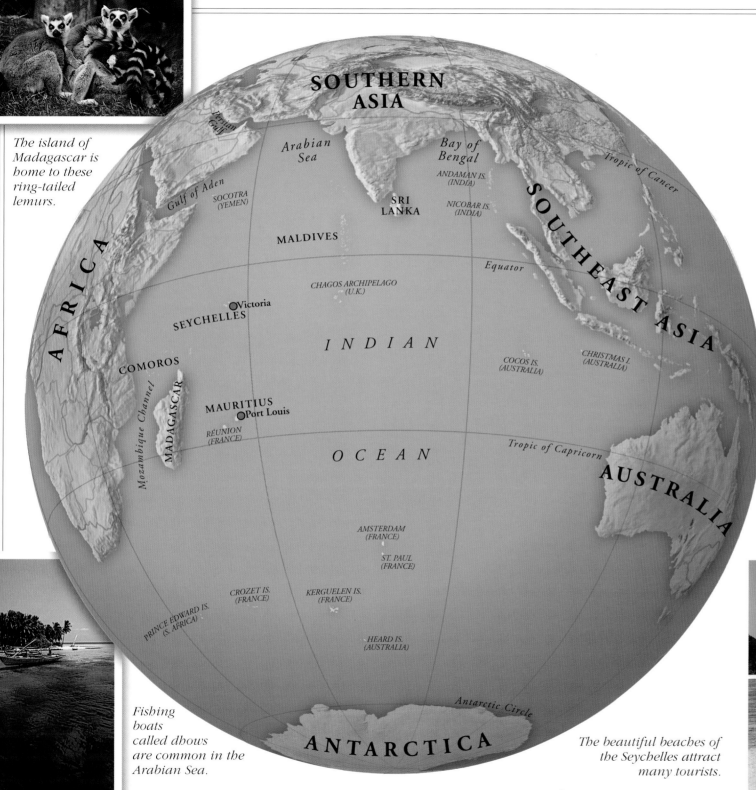

SOUTHERN ASIA

Persian Gulf

Arabian Sea

Bay of Bengal

Tropic of Cancer

Gulf of Aden

ANDAMAN IS. (INDIA)

SOCOTRA (YEMEN)

SRI LANKA

NICOBAR IS. (INDIA)

MALDIVES

SOUTHEAST ASIA

Equator

CHAGOS ARCHIPELAGO (U.K.)

AFRICA

Victoria

SEYCHELLES

I N D I A N

COMOROS

COCOS IS. (AUSTRALIA)

CHRISTMAS I. (AUSTRALIA)

Mozambique Channel

MAURITIUS

Port Louis

MADAGASCAR

RÉUNION (FRANCE)

O C E A N

Tropic of Capricorn

AUSTRALIA

AMSTERDAM (FRANCE)

ST. PAUL (FRANCE)

CROZET IS. (FRANCE)

KERGUELEN IS. (FRANCE)

PRINCE EDWARD IS. (S. AFRICA)

HEARD IS. (AUSTRALIA)

Antarctic Circle

ANTARCTICA

The island of Madagascar is home to these ring-tailed lemurs.

Fishing boats called dhows are common in the Arabian Sea.

The beautiful beaches of the Seychelles attract many tourists.

Indian Ocean

The Indian Ocean, the world's third largest ocean, extends from southern Asia to Antarctica. The northern part of the ocean contains busy shipping lanes. Huge tankers, for example, cross the ocean to transport oil from the Persian Gulf to many parts of the world. Fishing vessels also sail the ocean, especially off the west coast of India. But fish spoils quickly in the hot weather in the northern and central parts of the Indian Ocean.

Hidden beneath the waves are long ocean ridges and deep trenches. These features are the edges of plates that form the Earth's hard outer layers. Around 180 million years ago, India was located near Antarctica. But a plate carrying India broke away and moved north, colliding with Asia around 50 million years ago. These plate movements created the modern Indian Ocean.

The Arctic

The Arctic Ocean, the smallest of the world's four oceans, lies around the North Pole on the top of the world. Ice covers much of the ocean for most of the year. Ships carrying goods must be accompanied by icebreakers that cut paths through the ice.

The region called the Arctic also includes the northern parts of Asia, Europe, and North America surrounding the Arctic Ocean. Ice sheets cover some areas. Greenland, for example, is blanketed by the world's second largest ice sheet. However, in parts of the Arctic, the snow and ice melt in the summer and plants grow. Various peoples, such as the Inuit of North America, live in the Arctic. Their traditional way of life has depended mainly on hunting such animals as polar bears, and fishing.

The Hare people of the Canadian Arctic are so called because they make their clothes from hare skins.

A small iceberg at Ellesmere Island, in northern Canada.

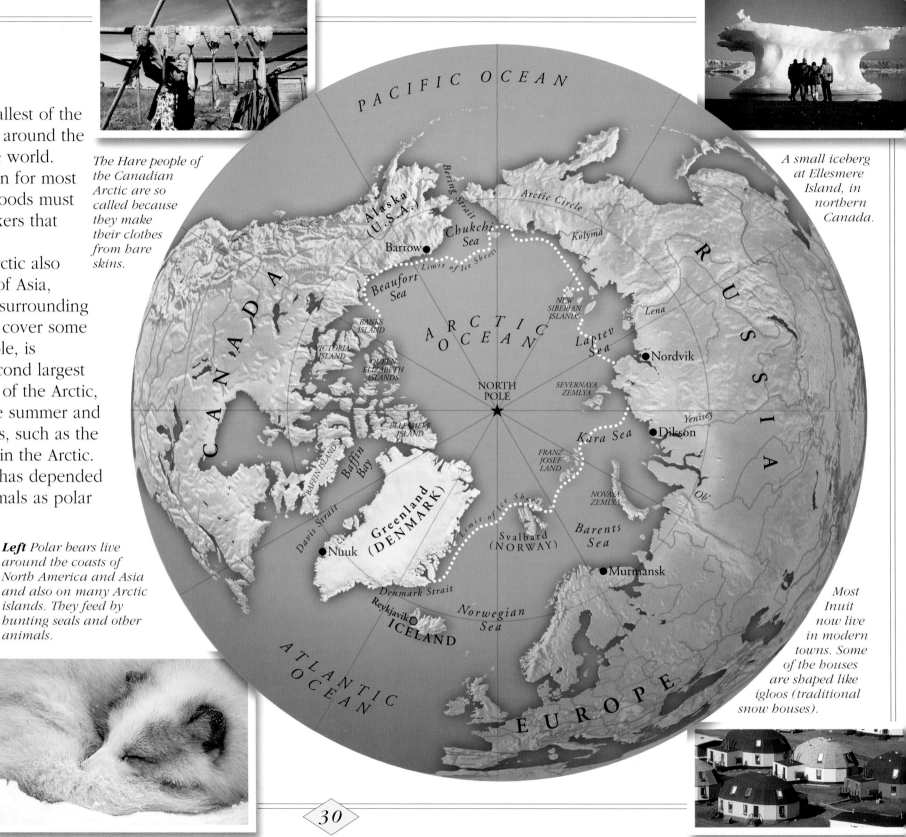

Left *Polar bears live around the coasts of North America and Asia and also on many Arctic islands. They feed by hunting seals and other animals.*

Right *The Arctic fox uses camouflage to help it catch its prey. Its coat changes from gray or brown in the summer to white in the winter, when snow covers the land.*

Most Inuit now live in modern towns. Some of the houses are shaped like igloos (traditional snow houses).

The Antarctic

Antarctica, the world's fifth largest continent, lies around the South Pole at the bottom of the globe. From space, you would see that it is mainly covered by the world's largest ice sheet, and surrounded by frozen seas. But you would also see high mountains jutting through the ice in places. In some parts of the continent, the ice is 15,700 feet (4,800 m) thick.

This frozen land is the world's coldest place. It is swept by strong winds that blow loose snow across the surface, causing blinding blizzards. Some scientists go there to study the continent and its weather, and some tourists now visit the continent. But no one lives in Antarctica all the time.

Leopard seals feed on young seals, penguins, and other creatures around Antarctica.

Map labels

ATLANTIC OCEAN

Scotia Sea

Antarctic Circle

SOUTH AMERICA

Drake Passage

ANTARCTIC PENINSULA

Weddell Sea

COATS LAND

QUEEN MAUD LAND

KERGUELEN IS. (FRANCE)

INDIAN OCEAN

BERKNER ISLAND

RONNE ICE SHELF

Lambert Glacier

Mackenzie Bay

Bellingshausen Sea

SOUTH POLE

TRANSANTARCTIC MOUNTAINS

Vinson Massif

Vostok Station (Russia)

MARIE BYRD LAND

WILKES LAND

ROSS ICE SHELF

McMurdo Station (U.S.A.)

ROOSEVELT ISLAND

Amundsen Sea

Mt. Erebus

Ross Sea

PACIFIC OCEAN

NEW ZEALAND

Penguins are flightless birds found in Antarctica.

Huge icebergs break away from the ice sheet of Antarctica and drift northward. About eight-ninths of the ice is hidden beneath the water.

Index

Picture Credits

Maps:
Mountain High Maps
©Digital Wisdom.

Illustrations:
Alastair Campbell;
Nicholas Rowland.

Photographs:
Peter Hince; Global
Scenes/National Tourist
Organization of Greece;
PhotoDisc, Inc.; Digital
Vision.

32